SIMPLYFING EXPRESSIONS

ALGEBRA

PRACTICE PROBLEMS

WITH ANSWER KEY AND

STEP BY STEP SOLUTIONS

Copyright © 2020 N. Hirn

All rights reserved.

Teachers and/or parents who purchase this workbook may make photocopies of the worksheets and solution sheet to use for instructional purposes only.

Reproducing this material for commercial purposes is not permitted.

HOW TO USE THIS WORKBOOK

The study of Mathematics requires understanding of the concepts taught as well as practicing what is learned. It is advisable to practice mathematical problems using pencil and paper to allow the student to follow their train of thoughts as they write each step for the solutions.

This workbook provides problems that require the knowledge and use of the SIMPLIFYING EXPRESSIONS rules. These rules are summarized in the next section as a refresher for students.

When working out each problem, it is important to learn, understand and apply the rules of SIMPLIFYING EXPRESSIONS in order to evaluate the mathematical expressions correctly.

Hints are provided when a problem requires the use of concepts that are not part of this workbook. Students should be familiar with some of these concepts.

The workbook is divided into five sections:

1. The first section includes a detail explanation of the topic discussed in the workbook with examples illustrating various problems and how each problem is being solved.

2. The second section includes a problem per page with ample lined space for the student to solve the problem. Using a pencil will allow the student to erase and redo steps as they become necessary.

3. The third section is the answer key.

4. The fourth section includes detailed solutions for each problem explaining the steps involved in reaching the answer.

5. The final section includes blank worksheets provided for the students to practice additional problems from the assigned text book.

In order to succeed the student should master the material in this section and practice solving each problem.

RULES FOR SIMPLIFYING EXPRESSIONS

When working on a mathematical expression it is important to understand how to evaluate each expression in order to obtain the correct answer.

In order to learn about Simplifying Expressions, it is important to first understand the following terms:

1. Arithmetic Properties
2. Terms
3. The Value of an Expression

This knowledge can then be used when learning to evaluate the rules associated with "Simplifying Expressions" and how to use them to solve a problem.

Explanation

1. Arithmetic Properties

The following is the list of all the arithmetic properties used in math. The rules governing each property will be discussed in detail with examples.

- **A. Commutative Property of Addition**
- **B. Commutative Property of Multiplication**
- **C. Associative Property of Addition**
- **D. Associative Property of Multiplication**
- **E. Distributive Property**
- **F. Additive Identity Property**
- **G. Multiplicative Identity Property**
- **H. Additive Inverse Property**
- **I. Multiplicative Inverse Property**

A. Commutative Property of Addition

In mathematics, when two numbers are added it makes no difference in which order the two numbers are placed. Changing the order of the numbers does not change the answer. This property is called:

"Commutative property of Addition".

To summarize the Commutative Property of Addition in symbols it can be written as follows: $a + b = b + a$

The Commutative property of Addition applies to addition problems only.

Example: $2 + 4 = 4 + 2$

B. Commutative Property of Multiplication

In mathematics, when two numbers are multiplied it makes no difference in which order the two numbers are placed. Changing the order of the numbers does not change the answer. This property is called:

"Commutative property of Multiplication".

To summarize the Commutative Property of Multiplication in symbols it can be written as follows: $ab = ba$ or $a(b) = b(a)$

The Commutative property of Multiplication applies to multiplication problems only.

Example: $2(4) = 4(2)$

Hint:

The Subtraction and Division operations are not Commutative. The order in which we subtract and divide makes a difference in the answer.

C. Associative Property of Addition

Associative operations have to do with grouping numbers.

When three numbers are added, it makes no difference which two are added first.

The answer will not change when the grouping or the association of the numbers are changed. This property is called:

"Associative Property of Addition"

To summarize the Associative Property of Addition in symbols it can be written as follows: $a + (b + c) = (a + b) + c$

The Associative property of addition applies to addition problems only.

Example:

$2 + 4 + 5 = 2 + (4 + 5)$ *Or*

$(2 + 4) + 5 = 2 + (4 + 5)$

The associative property of addition is used to simplify mathematical expressions that include both numbers and variables. This is shown in the example below:

Example: $2 + (4 + x) = (2 + 4) = x = 6 + x$

D. Associative Property of Multiplication

Associative operations have to do with grouping numbers.

When three numbers are multiplied, it makes no difference which two are added first.

The answer will not change when the grouping of the numbers is changed. This property is called:

"Associative Property of Multiplication"

To summarize the Associative Property of Addition in symbols it can be written as follows: $a(bc) = (ab)c$

The Associative property of multiplication applies to multiplication problems only.

Example: $2(4 \cdot 5) = (2 \cdot 4)5$

The associative property of multiplication is used to simplify mathematical expressions that include both numbers and variables. This is shown in the example below:

Example: $2(4x) = (2 \cdot 4)x = x = 8x$

E. Distributive Property

The distributive property involves both the addition and multiplication operations. It is used frequently in algebra.

To summarize the distributive property in symbols it can be written as follows: $a(b + c) = ab + ac$

The term "Distribute" means multiply whatever is outside the parenthesis by each term inside the parenthesis. Multiplication distributes over addition.

Example: $2(4 + 3) = 2(4) + 2(3) = 14$

The distributive property is used to simplify mathematical expressions that include both numbers and variables.

Example: $2(4 + x) = (2)(4) + (2)(x) = 8 + 2x$

The following properties are not used often. They state that the numbers "0" and "1" are special numbers. The properties explained below will show how these special numbers are used in mathematics.

F. Additive Identity Property

This property deals with the number "0" when used in the addition operation.

It states that when the special number "0" is used in an addition operation, it will preserve the identity of the expression it is added to.

Example: $2 + 0 = 2$ and $0 + 2 = 2$

The same applies to variables as follows: $x + 0 = x$ *and* $0 + x = x$

G. Multiplicative Identity Property

This property deals with the number **1** when used in the multiplication operation.

When the special number **1** is used in a multiplication operation, it will preserve the identity of the expression it is multiplied by.

Example: $2(1) = 2$ and $1(2) = 2$

The same applies to variables as follows: $x(1) = x$ and $1(x) = x$

H. Additive Inverse Property

This property states that when adding two opposite numbers, the answer is zero.

Example: $2 + (-2) = 0$

The same applies to variables as follows: $x + (-x) = 0$

I. Multiplicative Inverse Property

This property states that every real number such as x except for the number **0**, has a reciprocal that is in the form of $\frac{1}{x}$.

When multiplying a number by its reciprocal, the answer is always **1**.

Example: $2\left(\frac{1}{2}\right) = 1$

The same applies to variables as follows: $x\left(\frac{1}{x}\right) = 1$

The number 0 has no reciprocal. If a number is divided by 0 the answer will be "UNDEFINED".

Example: $\frac{15}{0} = undefined$

The same applies to variables as follows: $\frac{x}{0} = undefined$

2. Terms

A term is either a number, or a number and one or more variable that are multiplied together.

A number term is called a Constant term.

A number multiplied by one or more variable multiplied together is called a Variable term.

A number multiplied by a variable is called the Coefficient of the variable.

Example:

Constant terms: 2; 4; 100 .... etc

Variable terms: $2x$; $4y$; $100xy$ .... etc

In the variable terms above $2x$; $4y$; $100xy$, the numbers in front of each variable are called the Coefficient of the variable.

Like Terms

Like terms are identified as follows:

- Two or more constant numbers are considered Like Terms.
- Two or more terms that have the same variable with the same exponent are considered Like terms.
- Two or more terms with more than one variable care considered Like terms if the variables are exactly the same with the same exponent.

Example:

2; 4; 100 are constant terms and are Like Terms.

$2x$; $4x$; $100x$ are variable terms. They are Like Terms because they have the same variable x that is raised to the same exponent, in this case the exponent on the x variable is 1.

$2x^2$; $4x^2$; $100x^2$ are variable terms. They are Like Terms because they have the same variable x that is raised to the same exponent, in this case the exponent on the x variable is **2.**

$2xy$; $4xy$; $100xy$ are variable terms. They are Like Terms because they have the same variable xy that is raised to the same exponent, in this case the exponent on the xy variable is **1.**

$2x^2y$; $4x^2y$; $100x^2y$ are variable terms. They are Like Terms because they have the same variable x^2y that is raised to the same exponent, in this case the exponent on the x^2y variable is $2\ for\ x\ and\ 1\ for\ y.$

Like Terms can be added and subtracted.

Constant Like terms are added by adding the numbers.

Example: $2 + 3 = 5$ *Terms are Constant Like terms and can be added.*

Variable Like Terms can be added by adding the numbers (called coefficients) while leaving the variable intact with no change. This is accomplished by using the distributive property as seen in the example below.
Example: $3x + 2x$ and $3x - 2x$

Use the distributive property. Combine and add the numbers while keeping the variable the same. $(3 + 2)x = 5x$

Use the distributive property. Combine and subtract the numbers while keeping the variable the same. $(3 - 2)x = x$

Example: $3x^2 + 2x^2$ and $3x^2 - 2x^2$

Use the distributive property. Combine and add the numbers while keeping the variable the same. $(3 + 2)x^2 = 5x^2$

Use the distributive property. Combine and subtract the numbers while keeping the variable the same. $(3 - 2)x^2 = x^2$

Example: $3xy + 2xy$ and $3xy - 2xy$

Use the distributive property. Combine and add the numbers while keeping the variable the same. $(3 + 2)xy = 5xy$

Use the distributive property. Combine and add the numbers while keeping the variable the same. $(3 - 2)xy = xy$

Example: $3x^2y + 2x^2y$ and $3x^2y - 2x^2y$
Use the distributive property. Combine and add the numbers while keeping the variable the same. $(3 + 2)x^2y = 5x^2y$

Use the distributive property. Combine and add the numbers while keeping the variable the same. $(3 - 2)x^2y = x^2y$

Unlike Terms

Terms that have different variables or a variable and a number cannot be added.

Example: $2x + 3$

These are Unlike terms. The first term is a variable term. The second term is a constant term. They cannot be added. The expression is left as it is with no change.

Example: $2x + 3y$

These are Unlike terms. They are both variable terms but each has a different variable. They cannot be added. The expression is left as it is with no change.

3. **Value of an Expression**

A value of an expression is found when a certain value is assigned to a variable in an expression.

The number that is assigned to the variable is substituted in the expression. The operation is then carried out to find the value of the expression based on the number assigned to the variable.

$2x + 3$ is an expression. The value it has will depend on what number we assign to the variable x.

Example: *Evaluate the expression* $2x + 3$ *for* $x = 5$

In order to solve the problem, the number 5 is substitutes for the variable x in the above expression and the operation is then carried out as follows:

$2(5) + 3 = 10 + 3 = 13$

PROBLEM # 1

SOLVE: $3x + 4x$

YOUR WORK:

_____

_____

PROBLEM # 2

SOLVE: $2 + (3 + y)$

YOUR WORK:

_____

_____

PROBLEM # 3

SOLVE: $3x + 5 + 2x - 3$

YOUR WORK:

_____

_____

_____

_____

PROBLEM # 4

SOLVE: $-4x + 8 - 5x - 10$

YOUR WORK:

_____

_____

_____

_____

_____

PROBLEM # 5

SOLVE: $5x + 8 - x - 6$

YOUR WORK:

_____

_____

_____

_____

_____

_____

PROBLEM # 6

SOLVE: $4 + 3(2y + 5) + 8y$

YOUR WORK:

___

PROBLEM # 7

SOLVE: $3y + y + 5 + 2y + 1$

YOUR WORK:

PROBLEM # 8

SOLVE: $5(x-2) - (3x+4)$

YOUR WORK:

PROBLEM # 9

SOLVE: $2t + 3(1 + 6t) + 2$

YOUR WORK:

PROBLEM # 10

SOLVE: $7x + 4 + 6x + 3$

YOUR WORK:

PROBLEM # 11

SOLVE: $6(3x + 2y)$

YOUR WORK:

PROBLEM # 12

SOLVE: $18y - 10y + 6y$

YOUR WORK:

___

PROBLEM # 13

SOLVE: $\frac{1}{2}(3x + 6)$

YOUR WORK:

PROBLEM # 14

SOLVE: $5(3y)$

YOUR WORK:

_____

_____

PROBLEM # 15

SOLVE: $4(\frac{1}{4}a)$

YOUR WORK:

_____

_____

_____

PROBLEM # 16

SOLVE: $3a - 2a + 5a$

YOUR WORK:

_____

_____

_____

_____

PROBLEM # 17

SOLVE: $5(2x-1)+4$

YOUR WORK:

PROBLEM # 18

SOLVE: $3 + (2 + 7x)4$

YOUR WORK:

PROBLEM # 19

SOLVE: $7(2x + 4y + 6) + 10$

YOUR WORK:

PROBLEM # 20

SOLVE: $3 + 4(5a + 3) + 4a$

YOUR WORK:

PROBLEM # 21

SOLVE: $9(3x + 5y + 7) + 10$

YOUR WORK:

PROBLEM # 22

SOLVE: $3(x-2) + 2(x-3)$

YOUR WORK:

PROBLEM # 23

SOLVE: $x + 1 + x + 2 + x + 3$

YOUR WORK:

PROBLEM # 24

SOLVE: $8 - 3(4x - 2) + 5x$

YOUR WORK:

PROBLEM # 25

SOLVE: $(3x - 1) + (2x - 4) - (5x + 1)$

YOUR WORK:

PROBLEM # 26

SOLVE: $(6x^2 - 3x + 2) - (4x^2 + 2x - 5)$

YOUR WORK:

PROBLEM # 27

SOLVE: $x - 6[2x + 4(x - 5)]$

YOUR WORK:

PROBLEM # 28

SOLVE: $(x^3 - x) - (x^2 + x) + (x^3 - 3) - (x^2 + 1)$

YOUR WORK:

PROBLEM # 29

SOLVE: $-3[2x - 4(3x + 1)]$

YOUR WORK:

PROBLEM # 30

SOLVE: $3x - 1$ when $x = 2$

YOUR WORK:

PROBLEM # 31

SOLVE: $(5x^2 - 4x + 2) + (3x^2 + 9x - 6)$

YOUR WORK:

PROBLEM # 32

SOLVE: $-2x - y - 9$ when $x = -3$ and $y = 5$

YOUR WORK:

PROBLEM # 33

SOLVE: $2x - 5x + 1$ when $x = -5$

YOUR WORK:

PROBLEM # 34

SOLVE: $4x - 3[2 - (3x + 4)]$

YOUR WORK:

PROBLEM # 35

SOLVE: $(6x - 5) - (3x + 4)$

YOUR WORK:

PROBLEM # 36

SOLVE: $-8x^3 + 7x^2 - 6x + 5 + 10x^3 + 3x^2 - 2x - 6$

YOUR WORK:

PROBLEM # 37

SOLVE: $2x - 3 - 7x$ when $x = -5$

YOUR WORK:

PROBLEM # 38

SOLVE: $x^2 + 6xy + 9y^2$ when $x = -3$ and $y = 5$

YOUR WORK:

PROBLEM # 39

SOLVE: $4x^3(5x^2 - 3x + 1)$

YOUR WORK:

PROBLEM # 40

SOLVE: $x^2 + 2xy + y^2$ when $x = 2$ and $y = 3$

YOUR WORK:

PROBLEM # 41

SOLVE: $(x - y)^2$ when $x = -3$ and $y = 5$

YOUR WORK:

PROBLEM # 42

SOLVE: $(2x^3 + 5x^2 + 3) - (4x^2 - 2x - 7) - (6x^3 - 3x + 1)$

YOUR WORK:

PROBLEM # 43

SOLVE: $x^2 - 2xy + y^2$ when $x = 3$ and $y = -4$

YOUR WORK:

PROBLEM # 44

SOLVE: $(3x^2 - 4xy + 2y^2) - (4x^2 + 3xy + y^2)$

YOUR WORK:

PROBLEM # 45

SOLVE: $5x^3 - 3x^2 + 4x - 5$ when $x = 2$

YOUR WORK:

ANSWER KEY

1. $7x$

2. $5 + y$

3. $5x + 2$

4. $-9x - 2$

5. $4x + 2$

6. $14y + 19$

7. $6y + 6$

8. $2x - 14$

9. $20t + 5$

10. $13x + 7$

11. $18x + 12y$

12. $14y$

13. $\frac{3}{2}x + 3$

14. $15y$

15. a

16. $6a$

17. $10x - 1$

18. $28x + 11$

19. $14x + 28y + 52$

20. $24a + 15$

21. $27x + 45y + 73$

22. $5x - 12$

23. $3x + 6$

24. $-7x + 14$

25. -6

26. $2x^2 - 5x + 7$

27. $-35x + 120$

28. $2x^3 - 2x^2 - 2x - 4$

29. $30x + 12$

30. 5

31. $8x^2 + 5x - 4$

32. -8

33. 16

34. $13x + 6$

35. $3x - 9$

36. $2x^3 + 10x^2 - 8x - 1$

37. 22

38. 144

39. $20x^5 - 12x^4 + 4x^3$

40. 25

41. 64

42. $-4x^3 + x^2 + 5x + 9$

43. 49

44. $-x^2 - 7xy + y^2$

45. 31

PROBLEM # 1

SOLVE: $3x + 4x$

CORRECT ANSWER: $7x$

These are Like Terms. They can be added. Use the distributive property. Combine and add the numbers while keeping the variable the same.

$3x + 4x = (3 + 4)x = 7x$

PROBLEM # 2

SOLVE: $2 + (3 + y)$

CORRECT ANSWER: $5 + y$

The associative property of addition is used to simplify mathematical expressions that include both numbers and variables.

Group the numbers as follows: $(2 + 3) + y$

Perform the addition operation to solve the problem: $(2 + 3) + y = 5 + y$

Hint:

The constant term 5 in the above answer is not the same like the variable term y. Therefore, the two terms cannot be added.

PROBLEM # 3

SOLVE: $3x + 5 + 2x - 3$

CORRECT ANSWER: $5x + 2$

The above problem utilizes three properties. They are as follows:

- Commutative property of addition.
- Associative property of addition.
- Distributive property.

The steps are as follows:

Using the Commutative Property of Addition, change the order of the terms as follows: $3x + 2x + 5 - 3$

Using the Associative Property of Addition, group the like terms as follows:

$(3x + 2x) + (5 - 3)$

Using the Distributive Property distribute the variable term over the addition operation as follows: $(3 + 2)x + (5 - 3)$

Finally solve the problem by adding and subtracting the numbers as follows: $5x + 2$

Hint:

The variable term $5x$ in the above answer is not the same like the constant term 2. Therefore, the two terms cannot be added.

PROBLEM # 4

SOLVE: $-4x + 8 - 5x - 10$

CORRECT ANSWER: $-9x - 2$

The above problem utilizes three properties. They are as follows:

- Commutative property of addition.
- Associative property of addition.
- Distributive property.

The steps are as follows:

Using the Commutative Property of Addition, change the order of the numbers and variables as follows: $-4x - 5x + 8 - 10$

Using the Associative Property of Addition, group the like terms as follows:

$(-4x - 5x) + (8 - 10)$

Using the Distributive Property distribute the variable term over the addition operation as follows: $(-4 - 5)x + (8 - 10)$

Finally solve the problem by adding and subtracting the numbers as follows: $-9x - 2$

Hint:

The variable term $-9x$ in the above answer is not the same like the constant term -2. Therefore, the two terms cannot be added.

PROBLEM # 5

SOLVE: $5x + 8 - x - 6$

CORRECT ANSWER: $4x + 2$

The above problem utilizes three properties. They are as follows:

- Commutative property of addition.
- Associative property of addition.
- Distributive property.

The steps are as follows:

Using the Commutative Property of Addition, change the order of the numbers and variables as follows: $5x + 8 - x - 6$

Using the Associative Property of Addition, group the like terms as follows:

$(5x - x) + (8 - 6)$

Using the Distributive Property distribute the variable term over the addition operation. Note that taking the variable x out of the parenthesis leaves the coefficient 1 as follows: $(5 - 1)x + (8 - 6)$

Finally solve the problem by adding and subtracting the numbers as follows: $4x + 2$

Hint:

The variable term $4x$ in the above answer is not the same like the constant term 2. Therefore, the two terms cannot be added.

PROBLEM # 6

SOLVE: $4 + 3(2y + 5) + 8y$

CORRECT ANSWER: $14y + 19$

Use the distributive property to open the parenthesis. This is done by multiplying the term outside the parenthesis by each term inside the parenthesis as follows:

$4 + (3 \cdot 2y + 3 \cdot 5) + 8y = 4 + 6y + 15 + 8y$

The above problem utilizes three properties. They are as follows:

- Commutative property of addition.
- Associative property of addition.
- Distributive property.

The steps are as follows:

Using the Commutative Property of Addition, change the order of the numbers and variables as follows: $6y + 8y + 15 + 4$

Using the Associative Property of Addition, group the like terms as follows:

$(6y + 8y) + (15 + 4)$

Using the Distributive Property distribute the variable term over the addition operation as follows: $(6 + 8)y + (15 + 4)$

Finally solve the problem by adding and subtracting the numbers as follows: $14y + 19$

Hint:

The variable term $14y$ in the above answer is not the same like the constant term 19. Therefore, the two terms cannot be added.

PROBLEM # 7

SOLVE: $3y + y + 5 + 2y + 1$

CORRECT ANSWER: $6y + 6$

The above problem utilizes three properties. They are as follows:

- Commutative property of addition.
- Associative property of addition.
- Distributive property.

The steps are as follows:

Using the Commutative Property of Addition, change the order of the numbers and variables as follows: $3y + y + 2y + 5 + 1$

Using the Associative Property of Addition, group the like terms as follows:

$(3y + y + 2y) + (5 + 1)$

Using the Distributive Property, distribute the variable term over the addition operation. Note that taking the variable y out of the parenthesis leaves the coefficient 1 as follows: $(3 + 1 + 2)y + (5 + 1)$

Finally solve the problem by adding and subtracting the numbers as follows: $6y + 6$

Hint:

The variable term $6y$ in the above answer is not the same like the constant term 6. Therefore, the two terms cannot be added.

PROBLEM # 8

SOLVE: $5(x - 2) - (3x + 4)$

CORRECT ANSWER: $2x - 14$

Use the distributive property to remove the parenthesis.

Note that there is -1 outside the term $(3x + 4)$

Rewrite the problem as follows: $5(x - 2) - 1(3x + 4)$

Distribute and remove the parenthesis as follows:

$(5 \cdot x - 5 \cdot 2) - (1 \cdot 3x + 1 \cdot 4)$

$5x - 10 - 3x - 4$

$5x - 3x - 10 - 4$

Combine like terms by performing the appropriate operation to solve the problem: $2x - 14$

Hint:

The variable term $2x$ in the above answer is not the same like the constant term -14. Therefore, the two terms cannot be added.

PROBLEM # 9

SOLVE: $2t + 3(1 + 6t) + 2$

CORRECT ANSWER: $20t + 5$

Use the distributive property to open the parenthesis. This is done by multiplying the term outside the parenthesis by each term inside the parenthesis as follows:

$2t + (3 \cdot 1 + 3 \cdot 6t) + 2$

$2t + 3 + 18t + 2$

The above problem utilizes three properties. They are as follows:

- Commutative property of addition.
- Associative property of addition.
- Distributive property.

The steps are as follows:

Using the Commutative Property of Addition, change the order of the numbers and variables as follows: $2t + 18t + 3 + 2$

Using the Associative Property of Addition, group the like terms as follows:

$(2t + 18t) + (3 + 2)$

Using the "Distributive Property" distribute the variable term over the addition operation as follows: $(2 + 18)t + (3 + 2)$

Finally solve the problem by adding and subtracting the numbers as follows: $20t + 5$

Hint:

The variable term $20t$ in the above answer is not the same like the constant term 5. Therefore, the two terms cannot be added.

PROBLEM # 10

SOLVE: $7x + 4 + 6x + 3$

CORRECT ANSWER: $13x + 7$

The above problem utilizes three properties. They are as follows:

- Commutative property of addition.
- Associative property of addition.
- Distributive property.

The steps are as follows:

Using the Commutative Property of Addition, change the order of the terms as follows: $7x + 6x + 4 + 3$

Using the Associative Property of Addition, group the like terms as follows:

$(7x + 6x) + (4 + 3)$

Using the Distributive Property distribute the variable term over the addition operation as follows: $(7+6)x+(4+3)$

Finally solve the problem by adding and subtracting the numbers as follows: $13x+7$

Hint:

The variable term $13x$ in the above answer is not the same like the constant term 7. Therefore, the two terms cannot be added.

PROBLEM # 11

SOLVE: $6(3x+2y)$

CORRECT ANSWER: $18x+12y$

Use the distributive property to multiply the **6** by every term inside the parenthesis as follows:

$(6 \cdot 3x + 6 \cdot 2y)$

$18x+12y$

Hint:

The variable term $18x$ in the above answer is not the same like the variable term $12y$. Therefore, the two terms cannot be added.

PROBLEM # 12

SOLVE: $18y-10y+6y$

CORRECT ANSWER: $14y$

Use the distributive property as follows:

$(18-10+6)y$

Add and subtract the numbers to solve the problem: $14y$

PROBLEM # 13

SOLVE: $\frac{1}{2}(3x+6)$

CORRECT ANSWER: $\frac{3}{2}x+3$

Use the distributive property to multiply the $\frac{1}{2}$ by every term inside the parenthesis as follows: $\left(\frac{1}{2}\cdot 3x + \frac{1}{2}\cdot 6\right)$

The problem can be rewritten as follows: $\left(\frac{1}{2}\cdot\frac{3x}{1} + \frac{1}{2}\cdot\frac{6}{1}\right)$

Multiply the numerators and denominator to solve the problem: $\frac{3}{2}x + \frac{6}{2}$

Simplify the result: $\frac{3}{2}x + 3$

Hint:

The variable term $\frac{3}{2}x$ in the above answer is not the same like the constant term 3. Therefore, the two terms cannot be added.

PROBLEM # 14

SOLVE: $5(3y)$

CORRECT ANSWER: $15y$

This is a multiplication problem.

Multiply the numbers while keeping the variable the same to solve the problem: $(5\cdot 3)y = 15y$

PROBLEM # 15

SOLVE: $4(\frac{1}{4}a)$

CORRECT ANSWER: a

This is a multiplication problem.

Multiply the numbers while keeping the variable the same to solve the problem: $(\frac{4}{1} \cdot \frac{1}{4})a$

Multiply the numerators and denominator to solve the problem:

$\frac{4}{4}a = 1a$ or just a

PROBLEM # 16

SOLVE: $3a - 2a + 5a$

CORRECT ANSWER: $6a$

Use the distributive property as follows: $(3 - 2 + 5)a$

Add and subtract the numbers to solve the problem: $6a$

PROBLEM # 17

SOLVE: $5(2x - 1) + 4$

CORRECT ANSWER: $10x - 1$

Use the distributive property to multiply the 5 by every term inside the parenthesis as follows:

$(5 \cdot 2x - 5 \cdot 1) + 4$

$10x - 5 + 4$

Combine like terms by adding the numbers to solve the problem: $10x - 1$

Hint:

The variable term $10x$ in the above answer is not the same like the constant term -1. Therefore, the two terms cannot be added.

PROBLEM # 18

SOLVE: $3 + (2 + 7x)4$

CORRECT ANSWER: $28x + 11$

Use the distributive property to multiply the **4** by every term inside the parenthesis as follows:

$3 + (2 \cdot 4 + 7x \cdot 4)$

$3 + 8 + 28x$

Combine like terms by adding the numbers to solve the problem: $11 + 28x$

This can be rewritten as: $28x + 11$

Hint:

The variable term $28x$ in the above answer is not the same like the constant term 11. Therefore, the two terms cannot be added.

PROBLEM # 19

SOLVE: $7(2x + 4y + 6) + 10$

CORRECT ANSWER: $14x + 28y + 52$

Use the distributive property to multiply the **7** by every term inside the parenthesis as follows:

$(7 \cdot 2x + 7 \cdot 4y + 7 \cdot 6) + 10$

$14x + 28y + 42 + 10$

Combine like terms by adding the numbers to solve the problem:

$14x + 28y + 52$

Hint:

The variable term **14x** in the above answer is not the same like the variable term **28y** and not like the constant the constant term **52**. Therefore, the three terms cannot be added.

PROBLEM # 20

SOLVE: $3 + 4(5a + 3) + 4a$

CORRECT ANSWER: $24a + 15$

Use the distributive property to multiply the **4** by every term inside the parenthesis as follows:

$3 + (4 \cdot 5a + 4 \cdot 3) + 4a$

$3 + 20a + 12 + 4a$

The above problem utilizes three properties. They are as follows:

- Commutative property of addition.
- Associative property of addition.
- Distributive property.

The steps are as follows:

Using the Commutative Property of Addition, change the order of the terms as follows: $20a + 4a + 3 + 12$

Using the Associative Property of Addition, group the like terms as follows:

$(20a + 4a) + (3 + 12)$

Using the Distributive Property distribute the variable term over the addition operation as follows: $(20 + 4)a + (3 + 12)$

Finally solve the problem by adding and subtracting the numbers as follows: $24a + 15$

Hint:

*The variable term **24a** in the above answer is not the same like the constant term **15**. Therefore, the two terms cannot be added.*

PROBLEM # 21

SOLVE: $9(3x + 5y + 7) + 10$

CORRECT ANSWER: $27x + 45y + 73$

Use the distributive property to multiply the **9** by every term inside the parenthesis as follows:

$(9 \cdot 3x + 9 \cdot 5y + 9 \cdot 7) + 10$

$27x + 45y + 63 + 10$

Combine like terms by adding the numbers to solve the problem:

$27x + 45y + 73$

Hint:

*The variable term **27x** in the above answer is not the same like the variable term **45y** and not like the constant the constant term **73**. Therefore, the three terms cannot be added.*

PROBLEM # 22

SOLVE: $3(x - 2) + 2(x - 3)$

CORRECT ANSWER: $5x - 12$

Use the distributive property to multiply the **3** by every term inside the first parenthesis and the **2** by every term inside the second parenthesis as follows:

$(3 \cdot x - 3 \cdot 2) + (2 \cdot x - 2 \cdot 3)$

$3x - 6 + 2x - 6$

Using the Commutative Property of Addition, change the order of the terms as follows: $3x + 2x - 6 - 6$

Combine like terms by performing the necessary operations to solve the problem: $5x - 12$

Hint:

The variable term $5x$ in the above answer is not the same like the constant term -12. Therefore, the two terms cannot be added.

PROBLEM # 23

SOLVE: $x + 1 + x + 2 + x + 3$

CORRECT ANSWER: $3x + 6$

Using the Commutative Property of Addition, change the order of the terms as follows: $x + x + x + 1 + 2 + 3$

Combine like terms by performing the addition operations to solve the problem: $3x + 6$

Hint:

The variable term $3x$ in the above answer is not the same like the constant term 6. Therefore, the two terms cannot be added.

PROBLEM # 24

SOLVE: $8 - 3(4x - 2) + 5x$

CORRECT ANSWER: $-7x + 14$

Use the distributive property to multiply the 3 by every term inside the parenthesis as follows:

$8 - (3 \cdot 4x - 3 \cdot 2) + 5x$

$8 - 12x + 6 + 5x$

Using the Commutative Property of Addition, change the order of the terms as follows: $-12x + 5x + 8 + 6$

Combine like terms by performing the necessary operations to solve the problem: $-7x + 14$

Hint:

The variable term $-7x$ in the above answer is not the same like the constant term 14. Therefore, the two terms cannot be added.

PROBLEM # 25

SOLVE: $(3x - 1) + (2x - 4) - (5x + 1)$

CORRECT ANSWER: -6

Use the distributive property to open the parenthesis. There is the number $+1$ outside the first two parenthesis and the number -1 outside the last parenthesis.

Perform the multiplication operation and open the parenthesis as follows:

$3x - 1 + 2x - 4 - 5x - 1$

Group the Like terms as follows: $3x + 2x - 5x - 1 - 4 - 1$

Combine Like terms by performing the appropriate operation as follows:

$5x - 5x - 5 - 1$

Continue to combine Like terms to solve the problem:

$5x - 5x - 5 - 1 = -6$

PROBLEM # 26

SOLVE: $(6x^2 - 3x + 2) - (4x^2 + 2x - 5)$

CORRECT ANSWER: $2x^2 - 5x + 7$

Use the distributive property to open the parenthesis. There is the number +1 outside the first two parenthesis and the number −1 outside the last parenthesis.

Perform the multiplication operation and open the parenthesis as follows:

$6x^2 - 3x + 2 - 4x^2 - 2x + 5$

Group Like terms in descending order starting with the variable with the highest exponent as follows: $6x^2 - 4x^2 - 3x - 2x + 2 + 5$

Combine Like terms by performing the appropriate operation to solve the problem as follows: $2x^2 - 5x + 7$

Hint:

The variable term $2x^2$ in the above answer is not the same like the variable term $-5x$ and not like the constant the constant term 7. Therefore, the three terms cannot be added.

PROBLEM # 27

SOLVE: $x - 6[2x + 4(x - 5)]$

CORRECT ANSWER: $-35x + 120$

Use the distributive property to open the inner parenthesis first by multiplying the number 4 by each term inside the parenthesis as follows:

$x - 6[2x + (4 \cdot x - 4 \cdot 5)]$

$x - 6[2x + 4x - 20]$

Use the distributive property to open the brackets by multiplying the number 6 by each term inside the brackets as follows:

$x - [6 \cdot 2x + 6 \cdot 4x - 6 \cdot 20]$

$x - 12x - 24x + 120$

Combine Like terms by performing the appropriate operation to solve the problem as follows: $-35x + 120$

Hint:

The variable term $-35x$ in the above answer is not the same like the constant term 120. Therefore, the two terms cannot be added.

PROBLEM # 28

SOLVE: $(x^3 - x) - (x^2 + x) + (x^3 - 3) - (x^2 + 1)$

CORRECT ANSWER: $2x^3 - 2x^2 - 2x - 4$

Use the distributive property to open the parenthesis. There is the number $+1$ outside the first and third parenthesis and the number -1 outside the second and last parenthesis.

Perform the multiplication operation and open the parenthesis as follows:

$x^3 - x - x^2 - x + x^3 - 3 - x^2 - 1$

Group Like terms in descending order starting with the variable with the highest exponent as follows:

$x^3 + x^3 - x^2 - x^2 - x - x - 3 - 1$

Combine Like terms by performing the appropriate operation to solve the problem as follows:

$2x^3 - 2x^2 - 2x - 4$

Hint:

The variable term $2x^3$ in the above answer is not the same like the variable term $-2x^2$ and not like the variable term $-2x$ and is not like the constant term -4. Therefore, the four terms cannot be added.

PROBLEM # 29

SOLVE: $-3[2x - 4(3x + 1)]$

CORRECT ANSWER: $30x + 12$

Use the distributive property to open the inner parenthesis first by multiplying the number 4 by each term inside the parenthesis as follows:

$-3[2x - (4 \cdot 3x + 4 \cdot 1)]$

$-3[2x - 12x - 4]$

Combine the like terms inside the bracket to obtain: $-3[-10x - 4]$

Use the distributive property to open the bracket by multiplying the number 3 by each term inside the bracket as follows:

$-[3 \cdot -10x - 3 \cdot 4]$

$-[-30x - 12]$

Use the distributive property to open the bracket by multiplying the number -1 by each term inside the bracket to get the following result: $30x + 12$

Combine Like terms by performing the appropriate operation to solve the problem as follows: $30x + 12$

Hint:

The variable term $30x$ in the above answer is not the same like the constant term 12. Therefore, the two terms cannot be added.

PROBLEM # 30

SOLVE: $3x - 1$ when $x = 2$

CORRECT ANSWER: 5

A value of an expression is found when a variable in an expression has been replaced by a number.

In the problem the variable x is replaced by the number **2**. The appropriate operation is then carried out to solve the problem as follows:

$3(2) - 1$

$6 - 1 = 5$

PROBLEM # 31

SOLVE: $(5x^2 - 4x + 2) + (3x^2 + 9x - 6)$

CORRECT ANSWER: $8x^2 + 5x - 4$

Use the distributive property to open the bracket by multiplying the number **1** by each term inside the bracket to get the following result:

$5x^2 - 4x + 2 + 3x^2 + 9x - 6$

Group Like terms in descending order starting with the variable with the highest exponent as follows:

$5x^2 + 3x^2 - 4x + 9x + 2 - 6$

Combine Like terms by performing the appropriate operation to solve the problem as follows: $8x^2 + 5x - 4$

Hint:

The variable term $8x^2$ in the above answer is not the same like the variable term $5x$ and not like the constant the constant term -4. Therefore, the three terms cannot be added.

PROBLEM # 32

SOLVE: $-2x - y - 9$ when $x = -3$ and $y = 5$

CORRECT ANSWER: -8

A value of an expression is found when a variable in an expression has been replaced by a number.

In the problem the variable x is replaced by the number -3.

The variable y is replaced by the number 5.

The appropriate operation is then carried out to solve the problem as follows:

$-2(-3) - 5 - 9$

$6 - 5 - 9 = -8$

PROBLEM # 33

SOLVE: $2x - 5x + 1$ when $x = -5$

CORRECT ANSWER: 16

A value of an expression is found when a variable in an expression has been replaced by a number.

In the problem the variable x is replaced by the number -5.

The appropriate operation is then carried out to solve the problem as follows:

$2(-5) - 5(-5) + 1$

$-10 + 25 + 1 = 16$

PROBLEM # 34

SOLVE: $4x - 3[2 - (3x + 4)]$

CORRECT ANSWER: $13x + 6$

Use the distributive property to open the inner parenthesis. There is a -1 outside the inner parenthesis. Open the parenthesis as follows:

$4x - 3[2 - (1 \cdot 3x + 1 \cdot 4)]$

$4x - 3[2 - 3x - 4]$

Use the distributive property to open the bracket by multiplying the number 3 by each term inside the bracket as follows:

$4x - [3 \cdot 2 - 3 \cdot 3x - 3 \cdot 4]$

$4x - 6 + 9x + 12$

Group Like terms as follows: $4x + 9x - 6 + 12$

Combine Like terms by performing the appropriate operation and solve the problem: $13x + 6$

Hint:

The variable term $13x$ in the above answer is not the same like the constant term 6. Therefore, the two terms cannot be added.

PROBLEM # 35

SOLVE: $(6x - 5) - (3x + 4)$

CORRECT ANSWER: $3x - 9$

Use the distributive property to open the parenthesis. There is the number $+1$ outside the first parenthesis and the number -1 outside the second parenthesis.

Perform the multiplication operation and open the parenthesis as follows:
$6x - 5 - 3x - 4$

Using the Commutative Property of Addition, change the order of the terms as follows: $6x - 3x - 5 - 4$

Combine Like terms by performing the appropriate operation to solve the problem as follows: $3x - 9$

Hint:

The variable term $3x$ in the above answer is not the same like the constant term -9. Therefore, the two terms cannot be added.

PROBLEM # 36

SOLVE: $-8x^3 + 7x^2 - 6x + 5 + 10x^3 + 3x^2 - 2x - 6$

CORRECT ANSWER: $2x^3 + 10x^2 - 8x - 1$

Group Like terms in descending order starting with the variable with the highest exponent as follows:

$-8x^3 + 10x^3 + 7x^2 + 3x^2 - 6x - 2x + 5 - 6$

Combine Like terms by performing the appropriate operation to solve the problem as follows: $2x^3 + 10x^2 - 8x - 1$

Hint:

The variable term $2x^3$ in the above answer is not the same like the variable term $10x^2$ and not like the variable term $-8x$ and is not like the constant term -1. Therefore, the four terms cannot be added.

PROBLEM # 37

SOLVE: $2x - 3 - 7x$ when $x = -5$

CORRECT ANSWER: 22

A value of an expression is found when a variable in an expression has been replaced by a number.

In the problem the variable x is replaced by the number -5.

The appropriate operation is then carried out to solve the problem as follows:

$2(-5) - 3 - 7(-5)$
$-10 - 3 + 35 = 22$

PROBLEM # 38

SOLVE: $x^2 + 6xy + 9y^2$ when $x = -3$ and $y = 5$

CORRECT ANSWER: 144

A value of an expression is found when a variable in an expression has been replaced by a number.

In the problem the variable x is replaced by the number -3.

The variable y is replaced by the number 5.

The appropriate operation is then carried out to solve the problem as follows:

$(-3)^2 + 6(-3)(5) + 9(5)^2$

$9 - 90 + 9(25)$

$9 - 90 + 225 = 144$

PROBLEM # 39

SOLVE: $4x^3(5x^2 - 3x + 1)$

CORRECT ANSWER: $20x^5 - 12x^4 + 4x^3$

Use the distributive property to open the parenthesis by multiplying the term $4x^3$ by each term inside the parenthesis as follows:

$4x^3 \cdot 5x^2 - 4x^3 \cdot 3x + 4x^3 \cdot 1$

Hint:

When multiplying the similar variables with different exponents, the exponents are added as follows:

$20x^{3+2} - 12x^{3+1} + 4x^3$

$20x^5 - 12x^4 + 4x^3$

Hint:

The variable term $20x^5$ in the above answer is not the same like the variable term $-12x^4$ and not like the variable term $4x^3$. Therefore, the three terms cannot be added.

PROBLEM # 40

SOLVE: $x^2 + 2xy + y^2$ when $x = 2$ and $y = 3$

CORRECT ANSWER: 25

A value of an expression is found when a variable in an expression has been replaced by a number.

In the problem the variable x is replaced by the number 2.

The variable y is replaced by the number 3.

The appropriate operation is then carried out to solve the problem as follows:

$(2)^2 + 2(2)(3) + (3)^2$

$4 + 12 + 9 = 25$

PROBLEM # 41

SOLVE: $(x - y)^2$ when $x = -3$ and $y = 5$

CORRECT ANSWER: 64

A value of an expression is found when a variable in an expression has been replaced by a number.

In the problem the variable x is replaced by the number -3.

The variable y is replaced by the number 5.

The appropriate operation is then carried out to solve the problem as follows:

$(-3 - 5)^2$

$(-8)^2 = 64$

PROBLEM # 42

SOLVE: $(2x^3 + 5x^2 + 3) - (4x^2 - 2x - 7) - (6x^3 - 3x + 1)$

CORRECT ANSWER: $-4x^3 + x^2 + 5x + 9$

Use the distributive property to open the parenthesis. There is the number $+1$ outside the first parenthesis and the number -1 outside the second and third parenthesis.

Perform the multiplication operation and open the parenthesis as follows:

$2x^3 + 5x^2 + 3 - 4x^2 + 2x + 7 - 6x^3 + 3x - 1$

Group Like terms in descending order starting with the variable with the highest exponent as follows:

$2x^3 - 6x^3 + 5x^2 - 4x^2 + 2x + 3x + 3 + 7 - 1$

Combine Like terms by performing the appropriate operation to solve the problem as follows: $-4x^3 + x^2 + 5x + 9$

Hint:

The variable term $-4x^3$ in the above answer is not the same like the variable term $1x^2$ and not like the variable term $5x$ and is not like the constant term 9. Therefore, the four terms cannot be added.

PROBLEM # 43

SOLVE: $x^2 - 2xy + y^2$ when $x = 3$ and $y = -4$

CORRECT ANSWER: 49

A value of an expression is found when a variable in an expression has been replaced by a number.

In the problem the variable x is replaced by the number **3.**

The variable y is replaced by the number **-4.**

The appropriate operation is then carried out to solve the problem as follows:

$(3)^2 - 2(3)(-4) + (-4)^2$

$9 + 24 + 16 = 49$

PROBLEM # 44

SOLVE: $(3x^2 - 4xy + 2y^2) - (4x^2 + 3xy + y^2)$

CORRECT ANSWER: $-x^2 - 7xy + y^2$

Use the distributive property to open the parenthesis. There is the number $+1$ outside the first parenthesis and the number -1 outside the second and third parenthesis.

Perform the multiplication operation and open the parenthesis as follows:

$3x^2 - 4xy + 2y^2 - 4x^2 - 3xy - y^2$

Group Like terms in descending order starting with the variable with the highest exponent as follows:

$3x^2 - 4x^2 - 4xy - 3xy + 2y^2 - y^2$

Combine Like terms by performing the appropriate operation to solve the problem as follows: $-x^2 - 7xy + y^2$

Hint:

The variable term $-x^2$ in the above answer is not the same like the variable term $-7xy$ and not like the variable term y^2. Therefore, the three terms cannot be added.

PROBLEM # 45

SOLVE: $5x^3 - 3x^2 + 4x - 5$ when $x = 2$

CORRECT ANSWER: 31

A value of an expression is found when a variable in an expression has been replaced by a number.

In the problem the variable x is replaced by the number **2.**

The appropriate operation is then carried out to solve the problem as follows:

$5(2)^3 - 3(2)^2 + 4(2) - 5$

$5(8) - 3(4) + 8 - 5$

$40 - 12 + 8 - 5 = 31$

ABOUT THE AUTHOR

Najwa Hirn holds a Bachelor of Science degree with honors in Engineering Technology. She has been working with Mathematics for over 25 years both professionally and privately. She taught math for many years.

Najwa is passionate about helping students succeed in Mathematics. She prides herself in being able to simplify math concept for students and teach every them according to their levels. Her step-by-step approach to solving problems has helped many students understand concepts better. She does not eliminate a step no matter how simple it may be since eliminating steps is what confuses many students.

Najwa Hirn can be reached at:

Learningmathquick@gmail.com